Foto File #1

Krupp Protze

and Variants

Foto File #1 is a visual reference of the The L2H43 and L2H143 "Krupp-Protze" six-wheeled 6x4 German truck and artillery tractor heavily used in World War II.

AF861231

SABOT Publications

Foto File #1 Krupp Protze
© 2017 SABOT Publications LLC unless otherwise stated. All rights reserved. This publication may not be reproduced in part or in whole without written permission from the publisher, except in cases where quotations are needed for reviews.

SABOT Publications
941 Cottontail Lane
St. Charles, Missouri 63303
www.sabotpub.com
info@sabotpub.com
United States of America
ISBN: 978-1-947552-00-5

Special thanks to our friends for their valuable contributions to this book:
Chris Leeman
John Charvat
Mike Petty
and all of the Soldiers, Sailors, Marines and Airmen who serve
Thank you!

Chris Mrosko
Brett Avants
Pearce Browning
SABOT Publications LLC

Printed in the United States of America

Introduction

The L2 H43 and L2 H143 "Krupp-Protze" was a six-wheeled 6x4 German truck and artillery tractor produced between 1934 and 1941. It was powered by a 55 hp or, from 1936, 60 hp Krupp M 304 4-cylinder gasoline engine. Its main purpose was to tow artillery or anti0tank guns, such as the 37mm PaK 36, and transport motorized infantry. The vehicle also was used as a generator carrier for towing the anti-aircraft spotlight.

The Krupp Protze was used extensively by all branches of the German Armed Forces in France, Sicily, the Eastern Front and in North Africa. About 7,000 vehicles were built of all variants. The wide deployment of the Protzkraftwagen – motorised limber - led to the nickname "Krupp Protze". Many of these vehicles served until the end of the war.

This book contains prints of original photographs of the Krupp Protze that have never before been published. The book is broken down into three chapters: 1 - Standard Troop Carrier Kfz 70; 2 - 37mm PaK transport Kfz 69, and 3 - Searchlight generator transport Kfz 83. The last chapter has several photographs of the searchlight itself, the generator carried in the Protze, some up close showing detail and others showing the searchlight in use or being towed.

Part 1: Standard Configuration Kfz 70

2,5 atü.

WH-88320

25

WH-54394

WH - 98057

275

5/32

WL-70948

Part 2: PaK Gun Towing Limber Kfz 69

WH-24116

WH-74772

WH
117081

WH-100734

39
WH-39830

WH-74537

WH-260 019

60

GASTHOF LUKASHANSL.

Part 3: Search Light Generator Carrier Kfz 83

WL-70459